Written by Gregory Wrightstone
Illustrated by Thiago Hellinger
Edited by Rafaella Nascimento, PhD
Reviewed by Peter Ridd, PhD, Angela Wheeler, Daniel Nebert, MD, Joseph Coyle and Gordon Tomb

© Copyright 2024 CO_2 Coalition

This series of books was inspired by Bert Goodrich and his book *Polar Bears Sleep Well.*

Chloe is a clownfish. She lives with her famil
on the Great Barrier Reef of Australia.

Chloe and her family swim and live
mong the corals and sea anemones
(pronounced uh·neh·muh·neez) that
make up this huge reef.

One day at school, her class had a special guest teacher, Doctor Moray Eel, a professor from Cook University. He told the class that their home and families were in great danger because humans were causing dangerous climate change

Doctor Moray Eel said that humans are adding carbon dioxide (CO_2) to the air when using fuels like gasoline and coal.

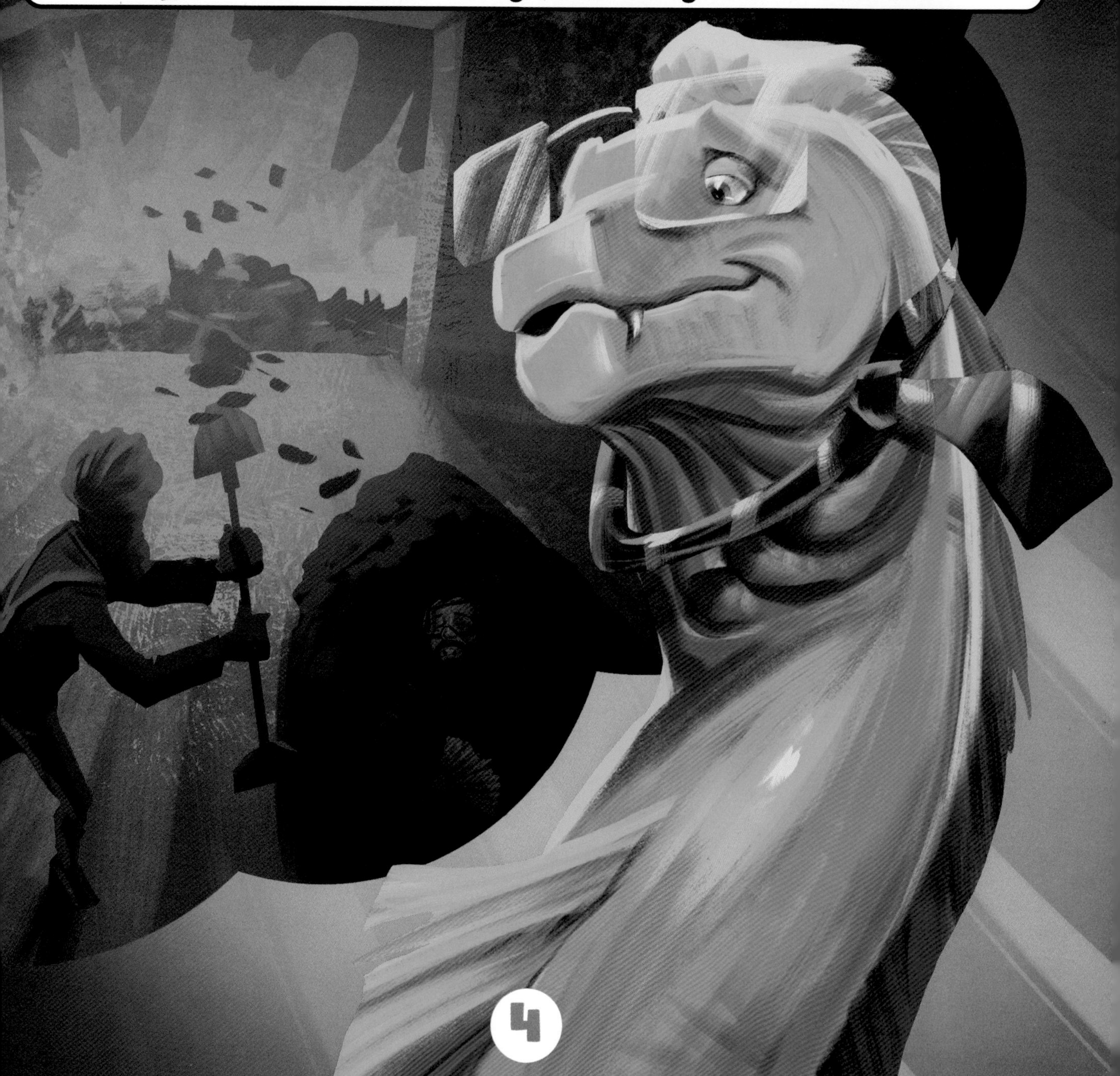

He said that this causes the atmosphere and the oceans to heat up because of a process known as the "greenhouse effect."

According to Doctor Moray Eel, carbon dioxide was also causing the ocean to become acidic, which will dissolve the reef and destroy it.

Chloe and her classmates were scared that they were in danger because of these human activities.

After Chloe returned home, she said:
Mama, I'm really worried about what Doctor Moray Eel told us today about climate change.
I can't stop thinking about it. If they are right, there will be no more clownfish and our family will be gone. Are they correct? Are we in danger?

Mama Clownfish was also concerned about this news.

I know that you're worried about this, and I know you have a hard time sleeping because of your worries.
I have to finish cooking dinner right now, and I have to bake seaweed cookies for the Reef Party. Try not to worry. We'll talk about this later.
CORALS

After dinner, Chloe and Mama sat down on their coral couch. Chloe had a troubled look on her face.

The next day, Mama asked Professor Maori to visit with their family after Chloe returned from school. She wanted to learn the truth from a reef expert.

Professor Maori, my daughter Chloe is very worried that our home here on the reef is being destroyed by climate change caused by humans.
There is a LOT that Doctor Moray Eel has wrong about our reef and our ocean.

When Chloe returned home from school that afternoon, she found her mother talking with Professor Maori.

Please come here, Chloe, and meet Professor Maori. He is here to discuss your concerns.
Pleased to meet you, Chloe.
Doctor Moray tells us that our ocean is heating up to dangerous levels and the increasing CO_2 is making our ocean acidic.
Happily, science tells us that this is completely incorrect. In fact, the Earth's climate has always been changing.

Chloe then asked if the coral and all of her friends who live on the reef will die if the ocean water gets a bit warmer.

No, Chloe. The ocean has warmed less than one degree (Centigrade) over the last 100 years and, according to scientists, it is about the same temperature that it was 300 years ago! [1]
The whole world was much hotter a few thousand years ago, and the reef was very healthy then.

Plus, corals like it hot and the hotter the better. Actually, most of the corals of the Great Barrier Reef live further south of what is called the Coral Triangle, on the equator, in waters around Indonesia and New Guinea, where the water is much warmer.

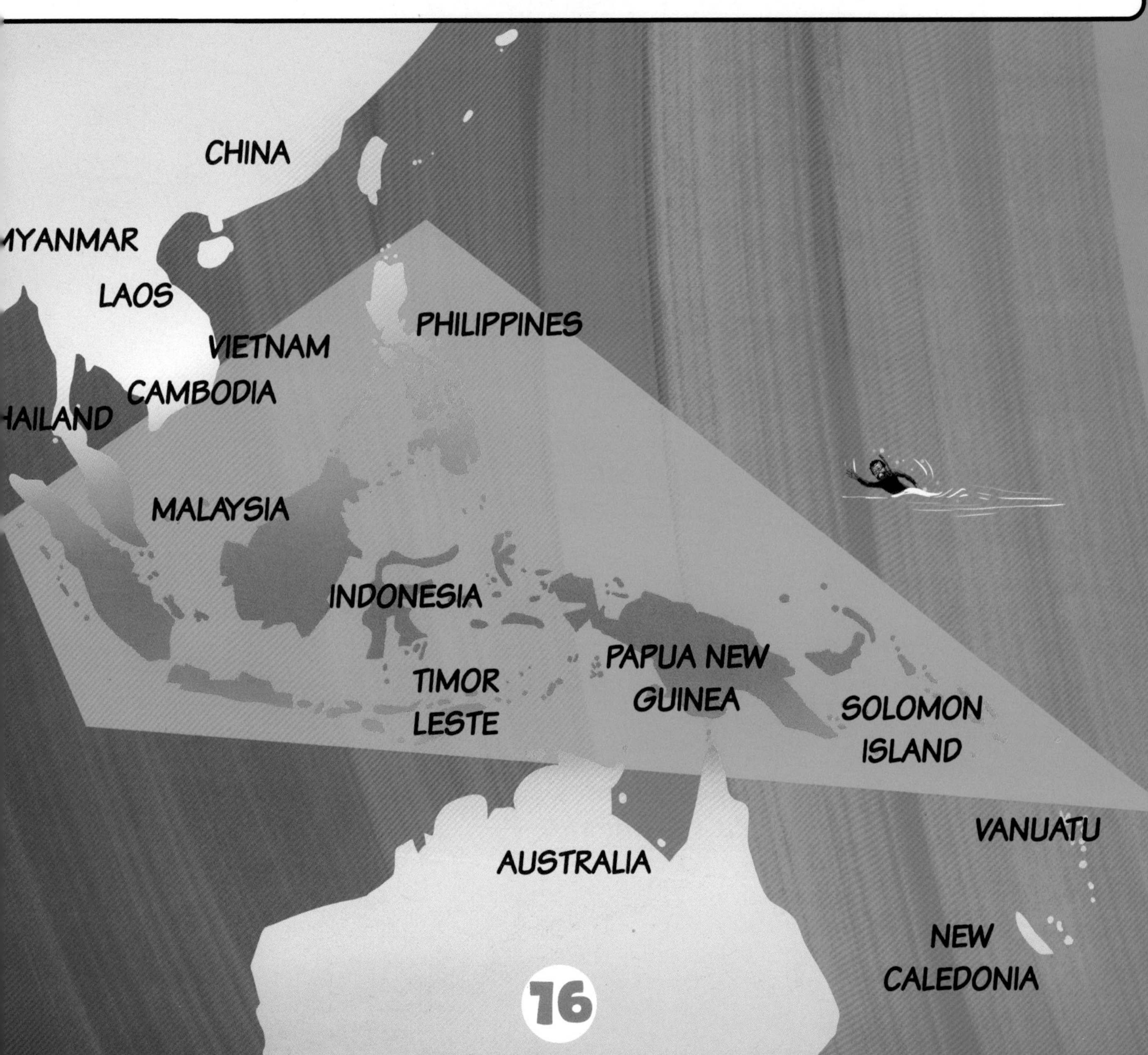

On top of that, the coral in the hotter Coral Triangle grow almost twice as fast as the corals of your home because the water there is warmer. You will be pleased to know that — since we began making surveys of the Great Barrier Reef in 1985 — the most recent surveys show the most growth yet.

That is certainly good news, Professor!
But what about the oceans becoming acidic?
Well, Chloe, your mother tells me that you learned about acids and bases just last week in your science class, so you know that liquids can be acidic or alkaline.
CORALS

If seawater ever became acidic, it could dissolve things like coral or seashells and that would be very bad for our reef home.
But Doctor Moray Eel is wrong, the sea water is NOT acidic and will not become acidic — even if humans burn more gasoline and coal.
In the past, CO_2 levels have been 20 times higher than today, but the ocean never became acidic.

Sleep well, Chloe.
Life for you on the Great Barrier Reef is good, our home is looking very healthy with record amounts of coral, and it's likely to get even better if the water warms a little bit more.

That night, after Mama tucked Chloe into bed, Chloe told her mother that she felt a lot better after learning the truth about climate change and healthy coral growth. As Mama left Chloe's room, she looked back and saw her little girl's eyes close as she drifted off to sleep.

And she was happy that her little girl was finally able to get some sleep, after learning the truth about their home on the Great Barrier.

his book and the CO_2 Learning Center were created by a group of distinguished xperts on the CO_2 Coalition's Education Committee. Along with CO_2 Coalition staff, these volunteer members include:

Jan Breslow, MD (Genetics)*

Sharon Camp, PhD (Chemistry)

Marty Cornell, BS (Chemistry)

John Droz, MS (Physics)

Bruce Everett, PhD (Economics)

Gordon Fulks, PhD (Physics)

William Happer, PhD (Physics)*

Hugh Kendrick, PhD (Nuclear Engineering)

Payne Kilbourn, (Certified Nuclear Engineer, NNPP)

Rafaella Nascimento, PhD (Chemistry)

Daniel Nebert, MD (Genetics)

Tom Sheahen, PhD (Physics)

Jim Steele, MS (Biology)

Mike Thompson, (Chief Meteorologist)

Gregory Wrightstone, MS (Geology)

Bob Zybach, PhD (Forest Ecology)

* Member, U.S. National Academy of Sciences

WHERE IS PETER RIDD?
I am snorkeling around throughout the book... You'll see me on each page if you take a close look!!

INFORMATION FOR PARENTS

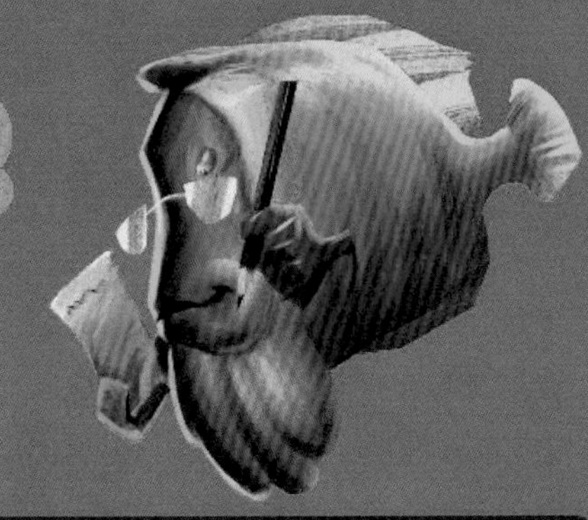

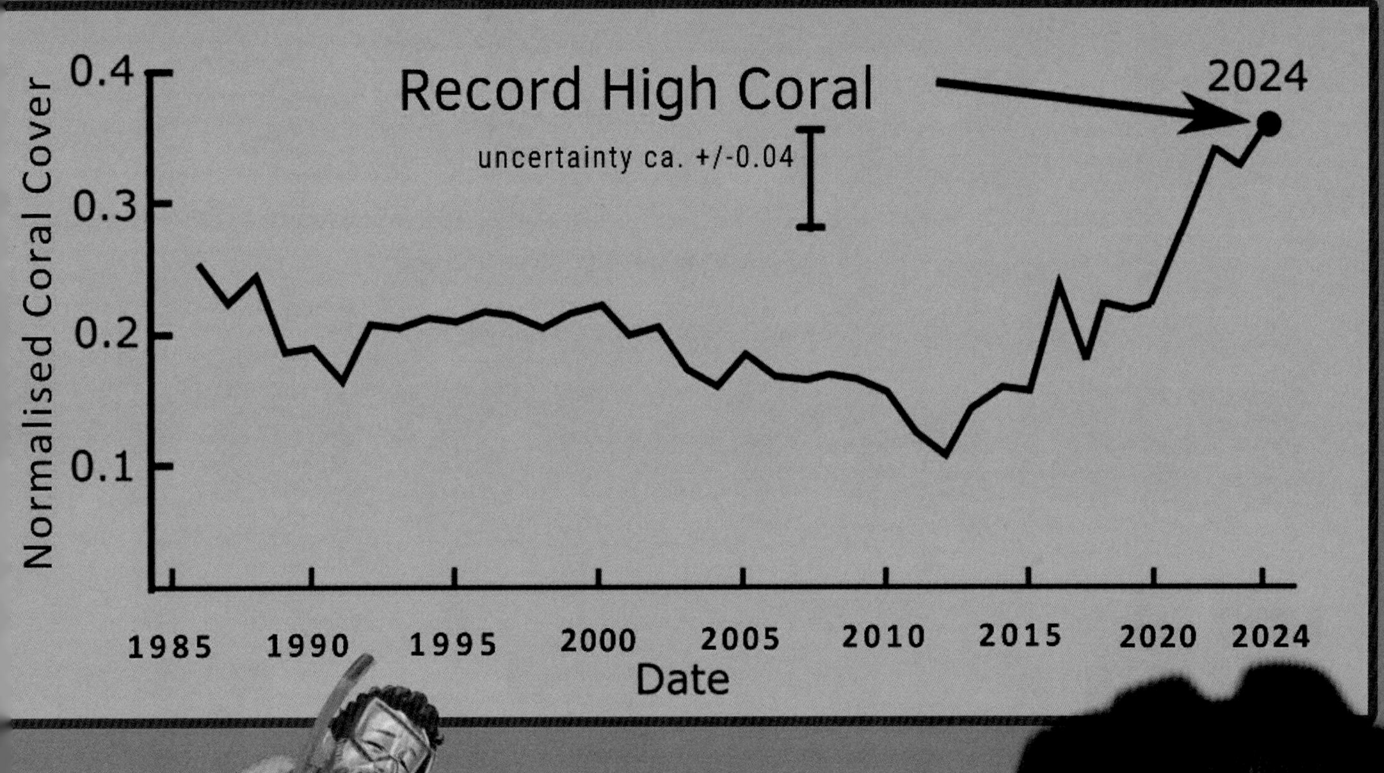

tps://www.aims.gov.au/research-topics/monitoring-and-discovery/monitoring-great-barrier-reef/reef-reports-hub

CLOWNFISH AND ANEMONES

Clownfish like Chloe and their relationship with anemones is a wonderful example of a symbiotic relationship in which two organisms benefit by living together.

Another amazing symbiosis is found in corals that have existed for 250 million years. Living where food is scarce, corals benefit by allowing algae to live inside their cells (polyps). The algae photosynthesize and make sugars that feed the coral. The coral's digestion of those sugars produces the CO_2 that algae need for photosynthesis. Corals and algae both benefit, as do the other plants and animals residing in coral reefs.

As it happens, different types of algae photosynthesize best in certain conditions depending on temperature, amount of sunlight or the water's saltiness. When a type of algae begins to underperform because of changing conditions, the coral ejects it to allow entry into its cell by another type of algae better adapted to the new environment And thus, a new symbiotic relationship is born.

It is the algae that provide the often stunning colors of the coral, so after the corals eject the old algae and before the new take up residence within the coral polyps, the coral whitens, or bleaches. Usually, once the new algae is in place, the coral's color is restored and the reef resumes development as a home for clownfish and numerous other creatures. Nature is truly amazing.

Note: Cyclones and starfish plagues are the primary threats to the Great Barrier Reef, which has withstood them for thousands of years.

MORE INFORMATION

1] Hendy, E.J., Lough, J.M. and Gagan, M.K. (2003). Historical mortality in nassive Porites from the central Great Barrier Reef, Australia: evidence for past environmental stress? Coral Reefs, 22(3), pp.207–215.

2] Lough, J.M. and Barnes, D.J. (2000). Environmental controls on growth of the massive coral Porites. Journal of Experimental Marine Biology and Ecology, 245(2), pp.225–243.

3] Ridd, P. (2023). Reef 'doomsayers' show breakdown in our science nstitutions. The Australian.

ttps://theconversation.com/coral-reefs-that-glow-bright-neon-during-bleaching-offer-hope-for-recovery-new-study-139048

LEARN MORE ABOUT CO_2, THE MIRACLE MOLECULE, AT CO2LEARNINGCENTER.COM

Hello friends!
Did you know this book has a lesson plan and activities? Just follow the QR Codes!

FUN FACTS

See how facts are fun at
CO2LearningCenter.com/fun-facts

VIDEOS
I love the awesome videos at the CO_2 Learning Center!
Me too! Science videos that are fun!

IN OUR NEXT BOOK, SPEND SOME TIME EXPLORING THE MALDIVES WITH FOXY THE FRUIT BAT!

WHAT IS CO_2 COALITION?

The CO_2 Coalition is a group of more than 180 world-class scientists, engineers and energy experts who study and report on the positive contribution of carbon dioxide to our ecosystems and humanity. The Coalition also educates policymakers and the general public on the true science of climate, promotes the discussion of the role of humans in the climate system, and explains the serious limitations of climate models while warning of the harmful consequences of mandated reductions in CO_2 emissions.

LEARN MORE AT CO2COALITION.ORG